T0132822

FLORA OF TROPICAL EAST AFRICA

FUMARIACEAE

G. Ll. Lucas

(East African Herbarium)

Annual or perennial herbs, erect, straggling or climbing ; sap watery. Leaves radical or alternate or more rarely subopposite, usually much dissected, sometimes ending in a branched tendril. Stipules absent. Flowers either in terminal or leaf-opposed racemes, sometimes spicate, rarely solitary, bisexual, usually irregular. Sepals 2, small, often caducous. Petals 4(–6 or more), if former then imbricate in 2 pairs and often connate at base, with 1 or both the outer pair spurred ; inner pair often united at their apex. Stamens either 4 free and opposite the petals, or 6 united into two groups of 3 by wing-like appendages of the filament to give a thin acuminate membrane surmounted by the 3 anthers. Ovary unilocular, superior, with 2 parietal placentas ; style filiform ; stigma lobed or entire. Ovules 1–many. Fruit a nutlet or capsule. Seeds 1–many.

Flowers yellow, rarely red or purple ; ovules numerous ; fruit
an elongate many seeded capsule . . . . . 1. **Corydalis**
Flowers white to pink ; ovule single ; fruit a ± globular nutlet 2. **Fumaria**

1. CORYDALIS

Vent., Choix Pl. : 19 (1803), *nom. conserv.*

Herbs, annual or perennial, with tap roots, tubers or rhizomes. Leaves alternate or rarely opposite, simple to variously divided, sometimes ending in a branched tendril. Inflorescence a panicle or raceme ; bracts persistent. Flowers irregular, yellow, rarely red or purple. Petals 4, with the upper petal spurred, free or sometimes united at their base. Stamens 6, in 2 bundles ; the outer anthers of each bundle are unilocular, the inner are bilocular ; the upper group also bears a nectary which is attached basally, lying in and partially attached to the inner side of the spurred petal. Stigma persistent, flattened and lobed ; style slender. Fruit a bicarpellate capsule having 2 valves. Seeds many, orbicular to reniform, smooth to pitted, bearing a caruncle.

A genus of over 300 species in the North Temperate region of both the Old and the New World, with but one species in tropical Africa. This belongs to Sect. *Corydalis*, subsect. *Ramoso-sibiricae* Fedde.

C. mildbraedii *Fedde* in F.R. 8 : 510 (1910) ; & in E. & P. Pf. 17b, ed. 2 : 133, fig. 69 G–J (1936). Type : Ruanda-Urundi, southern slopes of Mt. Karisimbi at 2500 m., *Mildbraed* (B, holo. !)

Glabrous usually glaucous perennial herb with a slender root. Stem up to 70 cm., much branched ; tendrils 0. Leaves bipinnatisect, characteristically multilobed ; apical lobes orbicular to broadly ovate ; lateral lobes

1

FIG. 1. *CORYDALIS MILDBRAEDII*—**1**, flowering branch, × 1 ; **2**, flower, × 4 ; **3**, upper and lateral petals and upper staminal bundle, × 4 ; **4**, lower petal, × 4 ; **5**, upper staminal bundle and nectary, × 4 ; **6**, lower staminal bundle and pistil, × 4 ; **7**, capsule, × 3 ; **8**, capsule after dehiscence, × 2 ; **9**, seed, × 6. 1, 7, from *Mienertzhagen in A.H.* 9298 ; 2–6, from *Tweedie* 1282 ; 8, from *Moore* 13 ; 9, from *Tweedie* 1522.

more spathulate to rounded ; petiole basally elongated, enlarged and dilated
to form a sheath which partly clasps the axis. Inflorescence a many-flowered
raceme ; flowers yellow ; bracts ± foliaceous, 2–5(–10) mm. long. Sepals
papery, rounded to subcordate, serrate. Corolla 1–1·5 cm. long ; petals in
2 pairs of 2, the outer pair distinctly keeled at apex, the inner pair narrowly
obovate, united at the apex. Ovary obovoid ; style curved towards tip.
Capsule laterally compressed, widened towards the rounded apex, pendulous,
about 12 mm. long, including the persistent 2·5 mm. long style and stigma,
and 3–5 mm. wide, dehiscing by the valves curling back from the apex
leaving the two placentas separate and exposed. Seeds brown-black,
reniform-orbicular, with small reticulately arranged pits. Fig. 1.

UGANDA. Kigezi District : Mt. Mgahinga, June 1949, *Purseglove* 2931 ! ; Mbale
   District : Elgon, Bulambuli, 4 Sept. 1932, *A. S. Thomas* 523 ! & 11 Nov. 1933,
   Tothill 2250 !
KENYA. Trans-Nzoia District : NE. Elgon, Dec. 1954, *Tweedie* 1282 ! & Mar. 1958,
   Tweedie 1522 ! ; Naivasha District : Kinangop, 21 Dec. 1930, *Napier* 720 ! ; North
   Nyeri District : Mt. Kenya, W. side above Naro Moru, 14 Dec. 1957, *Verdcourt*
   2057 ! ; Masai District : S. Mau Forest, Kepkogo, 4 Aug. 1956, *J. G. Williams in*
   E.A.H. 11079.
TANGANYIKA. Moshi District : western slopes of Kilimanjaro, Gararagua, 19 Apr.
   1957, *Greenway* 9181 !
DISTR. **U**2, 3 ; **K**3, 4, 6 ; **T**2 ; also in Congo Republic and Ruanda-Urundi (Virunga
   Mts.) and Ethiopia (SE.).
HAB. Upland rain-forest, bamboo-forest, upland moor & upland grassland, often at
   the marginal zone of the forest or near streams ; 2300–3300 m.

2. FUMARIA

L., Sp. Pl. : 699 (1753) & Gen. Pl., ed. 5 : 314 (1754) ; Pugsl. in
J.L.S. 44 : 233 (1919) & 47 : 427 (1927)

Herbs usually annual, erect to long-stemmed and straggling. Leaves
cauline, 2–4-pinnatisect ; tendrils absent ; sometimes the petiolule &
rhachis prehensile. Inflorescences in leaf-opposed racemes ; bracts usually
persistent. Flowers irregular, white to pink with some or all of the 4 petals
dark pink to purple at their apex. Sepals papery. Upper petal bears a
greenish keel which is often surrounded by a thinner margin or wing ; the
lower petal is never spurred and is ± spathulate. Stamens 6, united nor-
mally into 2 bundles of 3, as in *Corydalis*. Ovule 1. Fruit a nutlet, smooth
to rugose when dry.

A genus of over 50 species, mainly from southern Europe and North Africa to central
Asia and the Himalayas, with only one species in tropical Africa.

F. abyssinica *Hamm.* in Nov. Act. Soc. Uppsal., ser. 3, 2 : 275, t.6 (1857) ;
Pugsl. in J.L.S. 44 : 308 (1919). Type : Ethiopia, Semen, " Demerki ",
Schimper 1347 (B, holo.†, BM, K, iso. !)

Annual herb, in the early stages erect and tufted, usually becoming diffuse
and straggling, up to 60 cm. long. Root slender. Leaves longly petiolate,
occasionally the petiolule and rhachis prehensile ; lamina 2–3-pinnatisect,
glabrous ; leaflets oblong-lanceolate to linear-lanceolate, apices acute to
apiculate. Racemes (7–)11–15(–17)-flowered ; pedicels erect and spreading
in fruit, swollen beneath the fruit ; bracts linear, acute, varying from
nearly equal to longer than, the pedicel. Flowers 5–6·5 mm. long, pink with
a darker pink to purple tip. Sepals small, ovate-acuminate, irregularly
dentate, white to pale pink. Upper petal ± winged, sometimes flexed at
outer edge, keeled at apex, also having a short basal spur ; the inner petals
nearly straight, spathulate, united at their apex, and like the other petals
darker at their tips ; these 2 petals are also fluted, the middle fold being
the largest, with a single smaller fold either side; lower petal ± spathulate

FIG. 2. *FUMARIA ABYSSINICA*, from *Tweedie* 1709—**1**, part of flowering and fruiting plant, × 1 ; **2**, flower, × 10 ; **3**, upper and lateral petals and upper staminal bundle, × 10 ; **4**, lower petal, × 10 ; **5**, lower staminal bundle and pistil, × 10 ; **6**, nutlet, × 10.

with a small apical keel, free. Stamens united into 2 bundles of 3 with the upper unit bearing a small basal nectary, which hangs in and is partially attached to the spur of the upper petal. Fruit subrotund, obtuse to acute, ± rugose when dry, apiculate and slightly keeled with 2 poorly defined apical pits, only seen when mature and dry. Fig. 2.

UGANDA. Kigezi District : Muko, Lake Bunyoni, 30 Oct. 1929, *Snowden* 1626 ! & Kachwekano, June 1949, *Purseglove* 2908 ! ; Mbale District : Elgon, Bulambuli, 11 Nov. 1933, *Tothill* 2277 !
KENYA. Nakuru District : Lolderoto Escarpment, Oct. 1920, *Gardner* 2011 ! ; Kiambu District : Limuru, 26 Aug. 1940, *Greenway* 6004 ! ; Masai District : Ngong Hills, 5 July 1953, *Bally* 9009 !
TANGANYIKA. Mbulu District : Elanairobi Volcano, 20 Sept. 1932, *B. D. Burtt* 4179 ! ; Ufipa Plateau, Mtumba, 20 Dec. 1934, *Michelmore* 1079 ! ; Rungwe Mt., 4 Feb. 1914, *Stolz* 2503 !
DISTR. **U**2, 3 ; **K**3, 4, 6 ; **T**2, 4, 7 ; Congo Republic (Kivu), Ethiopia, Eritrea, Somaliland and Arabia
HAB. Upland rain-forest, bamboo forest & upland moor, becoming a weed of cultivations and more rarely of roadsides ; 1300–3200 m.
SYN.   *F. australis* Pugsl. in J.L.S. 44 : 309 (1919). Types : Tanganyika, Kilimanjaro, *Volkens* 953 & 1333 (BM, K, syn. !) ; Kenya, Nandi District, *Whyte* (K, syn. !)
VARIATION. The habit of *Fumaria* can vary considerably with the environment, mainly due to the influence of light-intensity and water-availability. The more exposed the habitat the smaller the leaflets and overall size of the plant, as with *Corydalis*. Also the earlier-formed leaflets are usually larger than the final ones. The colour of the flowers may also vary from pale pink to deep pink. Cleistogamous flowers are often to be found and these can cause confusion when identifying specimens and should therefore be disregarded.
   As a result of the additional gatherings from eastern Africa now available, many of the characters used by Pugsley to separate *F. australis* from *F. abyssinica*, have proved no longer to be of diagnostic value, as intermediate forms can now be found. For instance, the relative length of pedicel to bract can vary in one plant, and in the same inflorescence from base to apex, as well as between different plants, without any definite separation into two groups worthy of recognition. There is nothing in the known geographical range to suggest that more than one species is involved.

INDEX TO FUMARIACEAE